MYSTERIES OF THE MIND
Second Edition

How we think, what we believe, how we act, and why

Janusz Meyerhoff

CONTENTS

FOREWORD

Who are we?

This question has disturbed human minds through countless generations.

It's impossible to answer this question absolutely. Nevertheless, my intent is to describe human's characters as accurately as possible. I hope that the following explanations will serve for a better understanding of ourselves and what our real motives are, the way we act, think, feel, and why?

The mind is an indivisible part of a human being. However, to describe the mind I divided it artificially into various components, as I do not know any better way to make the subject of the mind clear.

SELF-CONSCIOUSNESS

Self-consciousness is a state of awareness of what happens inside the mind.
This is the most important step of the evolution of the mind and probably happened when our species emerged from the mist of the incipient state of being aware of the world, to the state of *self-consciousness*. How, when and why this took place, I will attempt to explain in the next chapter about the soul.

The same development can be observed in small children when they say "I" for the first time. It is the expression of *"self-consciousness"*, which will bring them many problems and also many accomplishments later in life.

This significant stage of the evolution of the mind is symbolically described in Genesis. Adam and Eve were expelled from an entirely instinctive "simple life" (paradise) because Adam ate the forbidden fruit (the apple – symbol of wisdom), thus losing his incorruptibility. However, he obtained the gift of being able to reach wisdom *(self-consciousness)*, through effort and suffering. This is the beginning of abstract thoughts, art, idols, *mythology*, beliefs, imagination, mental images, etc. – All mental characteristics that make us human!
However, none of these mental characteristics are immortal, which brings us to the immortal soul.

THE IMMORTAL SOUL

The definition of the *"soul"* in the Encyclopedia Britannica is: In religions and philosophies of the Western world, the concept of soul means immaterial aspect or the essence of a human being, which endorse individuality and is considered to be synonymous of the mind or ego or one-self. In theology, the soul is defined as being a part of the individual that he shares with divinity and which continues living after death."

These definitions leave more questions than they answers! What is the immaterial aspect of the soul? Is the soul made of thoughts, emotions, will, memory, intellect or consciousness? Is it a combination of all of these or only of one of them? All these questions have no answers!

Most probably, belief in soul first occurred when humans achieved awareness of individuality and the inevitability of death. When did this awareness happen? It's impossible to know. However, we do know that the first evidence of the belief in life after death was found in tomb offerings in archaeological excavations some of them were one hundred thousand years old. Ancient humans began to bury their deceased relations with objects, which according to their beliefs, they would need after death. That is considered to be an irrefutable proof of belief in the afterlife and consequently, of the immortality of the soul.

However, awareness of death was so terrifying that humans needed some ways to calm their fear. They believed they would not cease to exist at death, that their soul was immortal and eternal. This belief is the foundation of all religions. Almost all religions, with the exception of Buddhism, teach that one part of the soul is immortal and lives after death.

Before continuing, I want to mention a very strange experiment, which took place some time ago. A person was weighed before and immediately after dying. The result was astonishing! The person lost weight, very little, however, verifiable. That means that the soul which abandons the body during the process of dying possesses substance.

Nowadays, the word "soul" is used to explain the beliefs of other religions, both contemporary and ancient. That does not make any sense at all. The people who do not belong to our contemporary Western Civilization were thinking differently and had other concepts. The word "soul" has validity within present-day religions and philosophies of the Western Civilization only.

The number of different beliefs of humankind is absolutely astounding. My intention is to describe some of the various beliefs concerning the "soul" in different religions and the philosophies of the past and present, beginning with the oldest:

CHINCHORRO CULTURE

Until recently, the Chinchorro was an unknown culture. The first of their mummies was discovered on the West coast of Chile at the beginning of the twentieth century. When archeologists began to study these mummies they were surprised by their antiquity, as they dated from 5,000 BC to 1,700 BC. That proves that the Chinchorro were already mummifying their dead 2,000 years before the ancient Egyptians did and before the megalithic monoliths in Europe were built! Nevertheless, they were a primitive tribe in the hunter-gatherer stage of evolution. We do not know their beliefs, as they did not leave written records, but the sheer fact of mummifying is evidence of belief in the soul. They mummified everyone, from small children to elderly people, with no regard to age, sex, or social status.

ANCIENT EGYPT

In ancient Egypt, the symbol of the immortal element of the soul was the bird ("bai" or "ba") and the shadow ("ka"). This part was identical to the person to whom it belonged and could survive the body and resided in paintings or statues.

The KHNUM was the part of the soul that consumed the offerings and received the prayers of relations. The AKH was the spiritual part, obtained after death only. In ancient writings, we can see they believed that the soul was weighed after death and if it was lighter than a feather, then it could proceed and meet the Osiris (supreme God).

ANCIENT CHINESE

The ancient Chinese believed that the soul was divided after death into two parts: one went to paradise and the other remained in the body and had to be at peace. If not, it could become a devil. The tombs of important people, especially those of emperors, were decorated with ceremonial and practical objects that would serve them in the spiritual world after life.

HINDUISM

In the Hinduism, the concept of soul (Atman) means "one-self," and is the reflection of the absolute (Brahman). Soul, in the Hindu philosophy, is the only thing that exists and transmigrates from one body to another. The human life is cyclical: after dying the soul leaves the body and is born again in the body of another person, animal, or even a vegetable. This process is called *Samsara*. Reincarnation is determined by the actions of a person during this life and from his/her past lives.

THE EARLY HEBREWS

The early Hebrews had the concept that the body and soul are separate yet indivisible partners in human life. Rather than imprisoning or corrupting the soul, the body is a God-given tool for doing sacred work in the world. It requires protection, care and respect, because it is holy. Three words which over time developed the meaning of "soul" are present in Tanakh: *Neshemah Nefesh and Ruah*.

ZOROASTER

Zoroaster was the founder of the religion in ancient Persia. His teachings consisted of the belief that after death the soul of each person is judged in the "Bridge of

Discrimination." Those who passed judgment went to paradise, and those that failed went to hell.

BUDDHISM

Buddhism teaches that human existence consists of five realities (Skandhas): the material body, emotions, karmic perceptions, tendencies, and conscience. Each person is simply the temporary combination of these five realities and all of them are subject to continuous changes. Buddhists deny that these five realities can be considered soul (Atman), because they don't possess a permanent existence.

Buddha taught that all human existence is characterized by three realities: Anatman (no soul), Anitya (impermanence), and Dukkha (suffering). The doctrine of Atman includes belief of reincarnation: however, without transmigration of the soul.

GREEK PHILOSOPHERS

PHILO

Philo's teachings were strictly dualistic. The body is the prison of soul, which aspires to ascend to God. The ethics of Philo consisted of the liberation from instincts and declared the necessity of help from God, without which the human being can never be virtuous nor reach the truth. The soul of a wise man is free from bondage of instincts and is liberated. All other souls must pass to another body after death.

PYTHAGORAS

Pythagoras believed in the transmigration of the soul, from one body to another and even to the body of another species. If a man led a pure life his soul could be liberated from control of instincts. The soul consists of fire, air, heat, breath and had a great affinity with the celestial bodies.

Life consisted of a series of taboos and moral rules. Every night, the disciple had to ask himself three questions:

What did I fail to do today?

What good have I done?

What did I do that I should not have done?

PLATO

According to Plato the soul, in opposition to the matter, is the cause of spontaneous movement. A "soul of the world" is rational and good, and keeps order in the entire universe.

In the dialog "PHAEDO," Plato explains that the soul is a prisoner of the body and the slave of the senses.

He taught that philosophers must abstain from desires, pleasures, pains and fears.

The soul imagines that the senses are real, but they are not. Each pleasure and each pain are like nails that attach the body to the soul.

Plato also affirms that the soul exists after death and will be born again and again.

THE CELTS

The Celts inhabited the British Isles from the second century BC to the second century AD. They believed in the immortality of the soul, which after death was transferred into the body of a newborn child.

CHRISTIANITY

In Christianity, the soul either goes to paradise or to hell after death, depending on one's behavior. The soul must wait until the final judgment, which will arrive at the end of the world. In the purgatory, under the condition of good and correct behavior, the soul can be redeemed from sins. There is also a belief in the resurrection of the body and eternal life, but not in reincarnation.

TIBETAN BUDDHISM (Lamaism).

The beliefs of Lamaism are very similar to those of Buddhism - the nonexistence of soul.

The Tibetan Book of the Dead describes the road after death. Priests read a text in which they guide the deceased into an "intermediate state" (Bardo). "Real death" occurs three or four days after the first symptoms of death. Funerals continue for 49 days.

THE AZTECS

The Aztecs believed in the existence of the soul as well as in the afterlife. However, it was the way the Aztecs died, rather than the way they lived, that determined whether they would go to the sun god, or go to the dreary underworld. If a person died a normal death, his/her soul would have to pass through nine lives of the underworld before reaching Mictlan, the realm of the death. A warrior who died in battle or a woman who died in childbirth would go straight to the sun god.

THE ISLAM

In Islam, the soul is removed from the body by an angel of death at the moment of passing away. When the judgment day arrives Allah will decide who can enter paradise and who will descent to hell.

Every civilization developed its own belief for the path of soul after death.

However, each person that lived in these civilizations believed that only his belief was real (true), and that all the others were false.

Now, we will ask ourselves some fundamental questions:
Do all these beliefs are truth?
One of them?
None?
Perhaps, the truth is different for each person?
All these questions are difficult to reply and each person
must look for an answer himself.
How incredibly strong are all of these beliefs in existence
of the soul!
Religious beliefs were and are the cause of the
construction of extraordinary monuments: pyramids in
Egypt, cathedrals in Europe during the Middle Ages, the
unbelievably beautiful mausoleum Taj Mahal in India, etc.
All of them were built with the conviction that they
served the spiritual wellbeing of the whole community,
for the worship of gods and for the passage of souls to
the afterlife.

Nowadays, we also believe in god: however, the name of
our god is money. We also are building enormous
monuments to worship our new God: skyscrapers,
highways, jumbo planes, enormous cruise ships, and cars,
cars, and more cars everywhere.
However, everyone in our Western civilization is
convinced that only our way of life is real.

So, what is reality?

MENTAL IMAGES – REALITY OR ILLUSIONS?

How can we recognize reality? - Through "mental images" or ideas.

However, what are the "mental images?"

To understand this difficult subject, a profound change of mental attitude is necessary. To change the flow of thoughts, which are always set outside of ourselves and direct them inward - To observe what happens inside the mind and how the "mental images" arise. It is not an easy task!

We are not accustomed to this activity!

We know more about what happens in a distant galaxy or within an atom than what happens inside our heads!

The best way to understand the meaning of the "mental image" is through examples.

We can distinguish between two kinds of "mental images":

1) When we speak of a physical object, its "mental image" is similar to this object.

A chair is a physical object that we can experience through the senses. We can see it, smell it or touch it.

However, there is also a universal image of a chair, which has four legs, a seat, and a back, and which we have inside our minds.

Whenever we hear the word "chair" we immediately visualize a chair, not a particular one, but a "mental image" or an idea of a CHAIR.
In this case the chair is real: IT EXISTS!

2) However, when we speak of abstract "mental images"…it's different!
When we speak of a soul, ghost, aliens or about an angel, then every religion, every philosophy and each person has a different concept.
Abstract "mental images" cannot be experienced through the senses. We cannot see, hear, or smell them.
The one and only way is by "believing" in them.
We must understand that "mental images" exist only inside our mind!
The "mental images" are not real…It is our imagination!
* The next question is: Why, when, and how did we obtained the "mental images"?
When a man reached the consciousness of his own individuality, he found the world around him incomprehensible and threatening. He needed some help to control his fear.
Help appeared in the form of the "mental image"…Like so many times - just in time!
The abstract "mental images" of angels, saints, idols, etc. offers protection, consolation and defense against the threatening external world and horrifying idea of death.

The concept of "mental image" is represented metaphorically in various civilizations, and in different form.

In the Hindu philosophy, it's called MAYA, which means: illusion, image in the mirror, shadow or a fog on the road.

These metaphors imply that the world we are experiencing is not real, it is only an illusion!
Plato beautifully describes the same concept in his famous metaphor of the cave:

"Chained human beings are seated inside a cave in front of the wall and with the sun behind them. They can only see the shadows of their own silhouettes projected on the wall. They cannot see the sun behind them because they are prevented by the chain to which they are attached."

The symbol of reality is the sun: behind and invisible.
The chain is the symbol of the ignorance. We cannot see the sun (reality), because the chains are hindering to look behind and see the sun.
The shadows of the silhouettes projected on the wall shows the actions (bad or good).
However, it was not always like that.
The great spiritual teachers, the founders of religions, knew the dangers of living in illusion. They taught us: Do not make images of idols!
Now, what have we? Temples full of images and statues of gods, idols, or saints!

* We believe in what we like, what we wish, what we see on TV or what we read. However, we do not believe in what our reason tells us!
We live in a state of pure fantasy (unreality)!
There is a strict law: When our "mental image" differs from the reality, inexorably, the reality will prevail and the result will be appalling.
On the other hand, when our "mental image" is in agreement with the reality, whatever it is, we will receive

compensation and our life will be more pleasant and happy.

* In our Western civilization, there is only one therapy that gives a profound understanding of the mind. The psychoanalysis of Karl Jung, but it is not often applied. It is long process and does not always give the best results.

The last question:

* Does the reality exist outside of our minds?

This is a purely philosophical question and it's impossible to answer it here.

The subjects of "mental images," beliefs, and of reality are very ample and difficult. I am mentioning them here in a very simple and concise form.

Therefore, the next chapter will be about belief or reason?

BELIEFS OR REASON?

What does it mean to believe?
* Belief is a mental attitude of unconditional acceptance toward a proposition, without the full intellectual knowledge required to guarantee its truth.
* Reason, in philosophy, is the faculty or process of drawing logical inference.
* The conflict between these two opposite ways of thinking existed from the very beginnings of humanity. However, recently this has taken a drastic turn. For instance, the theory of evolution versus creation, which is so often and unnecessary discussed!
Exist two sorts of beliefs:
1) Religious beliefs in saints, angels, etc.
It is impossible to ascertain the veracity of religious beliefs or to understand them through logical thinking. We can only believe in them; there is no other way.
Believing in protective beings, like saints, guardian angels, etc. offers security and peace of mind and is a great help for people in distress – who is always happy?
However, many people enter into depression after a deception and start to doubt.
For example: If religion teaches that God is good and is all-powerful, the question arises, "why do we see so much evil in the world?"
If religions cannot provide an answer the door in the mind is opened to fear, doubt and disbelief! Sometimes, the belief in God can be used as an excuse to start wars

and persecute people belonging to other faiths. The priests, the preachers, the organizers of religions, the bishops, the cardinals, and even the butchers say, "God is with me." The businessman who takes advantage of his associate, and even the soldiers, who, during wartime kill their enemies – they all say, "God is with me."

Some religions affirm that only the people who share their beliefs can go to heaven. All non-believers will be condemned forever: therefore, the purpose of most religions is to convert as many people as possible to their particular faith and so to save them from eternal hell.
The consequences of these beliefs are fanaticism, wars, and the slaughter of millions of innocent people.

2) Belief in some event, news, or gossip makes no sense at all. Why do we need to believe, if we have reason (intelligence) to know?
These beliefs are based upon incessant repetitions of political propaganda, sensational news and commercial advertisement.

We believe that Princess D. was killed by Mr. X, who arranged the accident.

We believe, that the 9/11 terrorist attacks on the twin towers was a conspiracy.

We believe (and also buy) that the gadgets "X" are the best – only because we have seen them zillions of times on TV.

Instead of saying "I believe," we should be saying, "I have an opinion."

I personally have an opinion that it is foolish to "believe" in practically anything by repeating, like a parrot, information received from the communications media.

During my life, I have seen astonishingly large amount of "erroneous beliefs" about many ancient and recent historical events - all of them caused by brainwashing from different ways of communication.
This subject will be discussed in more detail in the next chapter.

FALSE CONCEPTS AND BELIEFS

Many beliefs and concepts that we take for granted as being absolutely true, are not true at all. Whenever we analyze current news, we discover that most information is biased or directly false.

Additionally, when we start to investigate history, both recent and ancient, from the scientific point of view, we discover with great surprise that many concepts are not based on historical truths.

How did we get all this misinformation?

The false concepts are the result of erroneous information released by the communications media, commercial advertisements, political propaganda, history's teachings, religious education, etc.

All of them tell people what and how they should think. It's a sad fact, but most people don't want to think for themselves!

Goebbels, the minister of propaganda of Nazi Germany, said, "The people will believe any lie if it is repeated many times."

* Commercial Advertisements

Industry and commerce must sell their products to make profit and for that they employ advertising agencies. These agencies use different tactics, all of them very harmful to human beings.

The first strategy is to confuse people utilizing flashes of a split-second duration and earsplitting music, which stupefy their clients and do not allow them to act reasonably and so to buy the superfluous, but desired gadget.

This is terribly harmful for all human beings, as it does not permit people to think clearly and consequently transforms humanity into a moron's society.

The second strategy is to increase desires by associating the object in the advertisement with an image of luxurious surroundings or of a beautiful and sexy girl.

This is the best strategy to sell everything - as all CEOs know well.

The influence of commercial advertisements in our daily lives is extremely harmful: take an advantage of the low instincts of human beings - greed, excessive desires - which accelerate the mind and destabilize society.

* Political propaganda.

Political propaganda is another technique of brain washing. The incessant repetition, especially during election times, of the names of political candidates by newspapers, radio and television, assures the victory of the candidate, who is neither appropriate nor truthful, but only spends more money (which he receives from rich and influential people). He gives promises that he cannot satisfy, once he is elected.

Consequently, the candidate becomes indebted to gain the election. However, when he wins the political position he must return money or favors and that is called CORRUPTION! Therefore, propaganda in the democratic system is the cause of corruption. It's as simple as that!

* The best option for everybody who wants to know the truth is to remember the bare facts only, and even those with caution.

Read the titles only, and forget all commentaries!

Sometime ago I asked a journalist, how can I know the truth?

He answered, "Buy all the newspapers and the truth must be somewhere in the middle"

I think this is a very long and tedious procedure!

* From here, I will proceed with the history.

The winners write the history! This proverb is very well-known. However, what does it mean? It means that almost everything that we learn from the history is NOT true. The archeological excavations solved many concepts that previously were considered fully proved, but resulted false.

We can only depend on books written by serious scientists, who were looking for the truth and not for fantasy.

I choose two examples only. The first 60 years ago. The second is from antiquity, 3,300 years ago.

I did not realize to what extent the history of World War II is distorted by the communications media, until a person in Argentina asked me if during World War II other people beside Jews died. He was convinced that only Jews died in concentrations camps and nobody else. I was so puzzled that I did not know what I should tell him.

The Holocaust was only one of many atrocities committed during this terrible war. It was a tragic episode, but did not have such historical importance as is actually attributed to it. The Nazis exterminated Jews, communists, homosexuals, the enemies of the Nazi regime, Gypsies, and non-Jewish people from almost every country in Europe.

However, the slaughter committed by Commissars, of the communist secret police (N.K.W.D, Cheka, GPU, KGB, etc.) upon the population of Russia, Poland, Ukraine, etc. is practically ignored by the communication media. Why?

It is a deliberate deformation of history! For history to be true it must be faithful to the importance of facts. It is important not to change one single fact into being unique and the most important.

The second example is from the Old Testament: many events are corroborated by archaeological discoveries, but many are NOT.
There is not a single shred of evidence of the Jews ever being in ancient Egypt. During two hundred years of excavations in Egypt only one inscription, the famous "Israel Estela" was found, in which Pharaoh Marneptah (1213 – 1204 BC) describes the rebellion in Palestine and mentions Israel as one of the tribes he defeated. However, this "Estela" is from a period after the date of the supposed Exodus. Also, no archeological proof of the Exodus in the Sinai Desert was ever found.
Even the highly respected Israeli archeologist Ze'ev Herzog of Tel Aviv University said, "The Israelites were never in Egypt and did not wander in the desert."
Nevertheless, the period around the IXX dynasty, in which the exodus was supposed to have taken place, is exceptionally well-known. Thousands of inscriptions were found on temples, tombs, palaces, etc. Now, we know

much more about ancient Egypt than about the historical events that happened much later.

These are only two examples of falsification of the history, but there are many, many more!

WHAT IS THE INTELLIGENCE?

There are several points of view on this topic. According to the Encyclopedia Britannica scientists differ on this subject, although the majority agrees on the importance of ability of adaptation to the environment which is the key to the understanding of what intelligence is and what purpose it serves.

* The process of adaptation involves:
1) Making changes in one-self to be better adapted to the environment.
2) Changing the environment to a more suitable one.
3) Moving to a different place.

Effective adaptation draws upon a number of cognitive processes, such as: perception, learning, memory, reasoning and solving problems.
The main trend in defining intelligence is that it is not only a cognitive or mental process, but rather a selective combination of all of these processes that is directed toward effective adaptation to the environment.
 Also, changes in the environmental conditions compel to adapt and teach a very important lesson: those who adapt progress, gain experience and intelligence.
Those who cannot adapt inexorably perish. However, when the change is too fast, the adaptation is difficult or directly impossible. In this case, extinction follows.

For example: When the dinosaurs disappeared 65 million years ago, the change in the environment was sudden and dramatic and because of its speed did not allow adaptation to the new environmental conditions. In this case, the change was caused by an impact of an asteroid *(another theory suggests that all dinosaurs died of boredom - no TV!)*This law of adaptation to the changes in environments is as valid now in the modern world as it was from the beginnings of life on Earth.

* Now, I will describe some examples of essential processes that are indispensable for the development of intelligence:

A) Understanding that "each cause produces an effect." At first look, it seems simple. However, it's not and most mistakes are linked to this simple law.

In addition, this law allows us to foresee the future and thereby to act appropriately.

B) The process of learning is another indispensable element of intelligence. It is the association between previous experience, retained in the memory, and the present situation.

This learning process ought to lead us not to repeat previous errors. The cat does not put its paws twice into the fire, but humans sometimes do!

C) Another essential component of intelligence is the ability to distinguish fantasy from reality. Our minds create dreams, fantasies, daydreams, illusions, etc. - which are not real.

The behavior must be based on reason and reality, and not on emotions, wishful thinking, or desires.

* From a genetics point of view, the latest scientific studies show that intelligence is in 50% of hereditary

origin (genes) and the other 50% is of the influence of the environment.

* From the neurological point of view, the intelligence is the capacity of the brain to process information simultaneously and in large number. The brain works in multiple forms and associates all the facts and draws the conclusions.
Therefore, the intelligence is the capacity to process multiple processes of reasoning simultaneously.

* Influence of society on the intelligence.
Unfortunately, the society hinders the development of intelligence. The incessant bombardment of communications media increases the interest in crime, the latest scandals, and the intimate life of celebrities in Hollywood and of sports superheroes.
Innate human curiosity is lost in brainless gossip. In addition, nationalism and religious fanaticism hinder free thought.
My sad conclusion is that society, in general, hinders the growth of intelligence. However, there is always a way out. The mind must not be necessarily conditioned by a society - it can free itself by developing curiosity and by always asking "why?"
In this way the intelligence will have the liberty to develop.

* How the intelligence evolved?
Our body and our mind evolved during the ice age in extremely harsh conditions. The ice age began some 110 thousand years ago and ended 10 thousand years ago. During 100 thousand years our ancestors had to cope with cold and ever changing environment and adapt

themselves to theses way of live. That is the reason that our minds evolved to be aggressive, competitive and intelligent.

THE INTELIGENCE IS THE TOOL OF THE INSTINCT OF SELF- PRESERVATION!

* Migration out of Africa

The migration of modern man from Africa had a big influence on his mental evolution.

Humans had to adapt themselves, physically and mentally, to different climates and diverse environments during the course of their wanderings.

Their digestive system adapted to eat practically everything: vegetables, roots, meat, rotten meat, fish, which gave them a big advantage.

Each new region, in which our ancestors arrived, had different climatic conditions: it's the main reason that we now have such an enormous diversity of cultures, languages, customs, beliefs, and mentalities.

Now this diversity is disappearing because of technological progress: international business, travels, tourism, airplanes, telephone, Internet, etc.

* Psychological evolution.

The new science of psychological evolution maintains that many characteristics of the mind can be explained by the instinct of procreation, which is more or less the same as what I have written about the instinct of self-preservation. However, I suppose, that this extremely rapid evolution was possible by the development of both instincts.

Psychologist, Miler affirms that the soul, conscience, and intelligence appeared by chance and their purpose was to look for a mate. The primitive females preferred the creativity instead of brute force and muscles. The consequence of this preference was that more babies were born with genes with superior intelligence. This process began 2.5 million years ago, when around the fire inside the caves, the primitive men tried to conquer the females using methods like: intelligence, humor, knowledge, creativity, etc. - everything to call the attention and conquer the female. This competition, between the males, was a principal cause of the evolution of the mind. The male who caught the attention of the female, had more descendants, all of them more intelligent and with superior genes.

In still older societies, the females preferred brute force and muscles, because that was more efficient for defense or attack.

Following the logic of Miler, the intellect has the same purpose as the crest of a rooster. It serves to catch the attention of the female and indicates that the man, or rooster, is healthy and it is a good material for reproduction.

* What is IQ?

In my opinion, the methods for the measurement of the intelligence quotient, IQ, is imprecise. A very important factor for the measurement of IQ is speed. However, speed leads to stress and mental acceleration, factors that prevent concentration, which is essential to knowledge. Therefore, the process of measuring intelligence prevents its exactitude.

At present, the measurements of IQ show a high growth during the last decades.

Are we really more intelligent today than before or only faster and more accelerated?

* By making use of intelligence we should achieve UNDERSTANDING.
This process seems simple;: however, it is not. For a true understanding a complete freedom is a must, without attachments of any kind. There can be no pressures or indoctrinations.
UNDERSTANDING can never be imposed from outside: each person must achieve it on his own.

HAPPINESS – ALWAYS BEYOND OUR REACH

Everybody wants to be happy. We go to the moon, buy the latest car, marry, murder or start a war – everything to reach happiness. However, we can never reach absolute happiness - it always eludes us.

What is happiness and what purpose does it serve?

From the point of view of the evolution, happiness is a process that preserves the genes of the organism to maintain the characteristics of the body and the mind until the next generation.

Biologists, speaking metaphorically, say that natural selection programs the characteristics of humans (body and mind) for specific tasks: intestines to digest, brains to think, ovaries to produce eggs, etc. and happiness serves to use these intestines, brains, ovaries, etc.

People eat, think, and have sex because by doing that, the body produces neuro-chemicals that make them happy. In addition, the reason that those neuro-chemicals are part of our inheritance is because they are doing well for themselves. Compare the destiny of these genes with the genes that would give a repugnant sensation after eating or after having sex.

The general principle is that genes that produce pleasure maintain, through countless generations, the same genes that are with us today. Therefore, the laws that cause happiness are planned not for our welfare, but for our long-term survival.

Think - this fact alone can produce unhappiness!

However, not only eating or having sex cause happiness. To have children, be successful at work, make friends or just to help somebody can cause happiness and consequently it's good for the genes.

The bad news is that happiness that can be acquired through numerous actions can also disappear. If happiness of a human being achieved by having sex, could last his entire life, he would make love only once. If happiness would last forever, we would be in the same situation as a drug dealer – he would sell drugs only once. Unfortunately, humans cannot understand this simple truth.

Therefore, the desire for happiness works far better under the illusion that we can achieve it. We are deluded into thinking that we will be truly happy with a new car, a new house or a new wife. Our false expectation is that happiness will remain and we will be truly happy and able to relax for the rest of our lives.

The fact remains, that things that were good for evolution are also good for the genes. Desires for food, sex, or social status are addictive - they bring pleasures that disappear, like the sensation of hunger that returns a few hours after eating.

Occasionally, even the happiest people sometimes can feel depressed, as also unhappy people can have some moments of happiness.

Our mentality is curious and logical: therefore, scientists invented several methods to calculate the degree of happiness.

How happy are we?

The degree of satisfaction within our own lives, naturally answered by the subjects of the study – as there is no other more objective way - consists of asking how many days you feel happy and how many days unhappy. When the amount of days of happiness is bigger, then your coefficient is higher and vice versa. Naturally, under the supposition that happiness, like goals in football, can be measured, and that the scientists that are making these measurements do not come from the same countries, neither speak the same language - otherwise it will leave a great margin for error.

It is almost impossible for the inhabitants of rich countries to understand the feeling of a person who lives in extreme poverty, like surviving on the streets of Calcutta without access to sufficient food or clean water.

From the perspective of a person that lives comfortably in a rich country, these multitudes and diseases seem an unimaginable suffering – hell!

However, not as unimaginable as a study that uses a method of calculation to research human happiness on a scale from one to seven. The poor inhabitants of streets in Calcutta marked a four - that is much more than one would imagine.

All these studies have established that the degree of happiness does not depend on factors like social status, wealth, education, or age. That is surprising for everyone who associates happiness with the possession of a TV or the latest car.

However, the diverse studies have also established that people who live in well-organized countries, with little corruption are happier than those that live in chaos and uncertainty.

However, it is not always like that - South Americans are generally happier than Asians and both of them live in corrupt and disorganized nations.

In addition, the studies have demonstrated that half of our happiness can be attributed to genetic influences (inherited - of "something within us"), and the other half to the influence of environment.

The big question is:

What is "something within us?"

Is it something that we inherited (genes), or something that we acquired?

So, is this "something" also partly hereditary?

All studies by psychologists, doctors, and scientists give ideas as to what to do to be happier and draw a rather obvious conclusion:

Be thankful to fellow beings –

Practice kindness –

Have a big family –

Have a good social life –

Smile –

Take care of the body -

Control stress -

In addition, an important part of being happy is to be involved in something - it does not matter what: hobbies, projects for future, trips, etc.

All that was nothing new for Buddha, who twenty-five hundred year ago, discovered the origin of this problem and based his philosophy on the existence of suffering on the conclusion that the cause of suffering is desire.

In spite of a very wrong opinion of this "strange" religion, it is a very optimistic philosophy. It teaches that it is possible to reach full happiness only by following the rules of Buddha, which consists of four truths and eight ways.

This subject is very long and I cannot expand more on it.

Buddhism is pragmatic and logical, and deals with nothing else than happiness. What is the purpose to know what is a soul, or God, or hell, if by knowing that we will never reach happiness?

In a famous metaphor, Buddha relates that a man was injured by an arrow.
What must he do?
Find the man who shot the arrow?
Inquire how old is he?
Who made the arrow?
Or remove the arrow from the body?
Buddhism, it's a strange paradox, is the only religion of Indo-European origin. It is surprising, but it has some similarity to the Epicurean school of ancient Greece. Perhaps this is not so surprising, if we take into account that they were distant cousins of European origin. All other religions, Judaism, Christianity, and Islam are of Semitic origin.
The original Buddhists teachings are (or rather were) based on reason only. All other religions teach beliefs and dogma.
The doctrine of Buddhism was originally "a way of life," and did not have anything to do with religion. Later on, it grew to be more complicated and Buddhism was divided into several religious sects.
 Now, according to differing points of view, this is good or bad news.
If everybody would attain happiness, then probably, this will be the last generation of the human species.
According to Darwin's laws of evolution, every species must be miserable to be able to survive and cannot live in a state of "eternal bliss." Nature does not sustain stagnation.
Everything static and do not adapt dies!

VIOLENCE

Violence is by far the most important characteristic of humankind: it shapes the destiny and it's indispensable for the survival of our species.

However, what purpose does it serve?

Violence serves to subjugate another person to obtain personal benefit and to satisfy desire.

* However, "Modern Man" has some special mental characteristics which distinguish him from the rest of all living species. The flesh-eating animals eat to kill. However, the human being kills for pleasure or for some noble ideal like: freedom, nationalism, religion, or any other ideal during never-ending wars.

In fact, researchers say that humans seem to crave violence just like they crave for sex, or drugs.

It is unquestionable that all predators must be violent. It is their nature. In order to survive they must kill their prey.

However, human violence goes far beyond that.

We are violent without apparent reason.

We are cruel to incredible limits: enjoy the bloody spectacles of death and suffering - bullfights, animals hunting, boxing, and sports in which the participants risks their lives.

The history of the "Homo-Sapiens-Sapiens" (us) is full of violence, wars, and exterminations.

There are no other species of "Homo" related to us.

WHY?

Scientists recently discovered that during the evolution of the "Modern Man," have cohabited with different species of the "Homo".

* The "Modern Man" arrived at the European continent around 40,000 years ago, and had to share the territory with the species of "Homo Neanderthal," (Our distant cousins) which lived on this continent for more than 200,000 years. Nevertheless, 30,000 years ago the "Homo Neanderthal" disappeared.

WHY? Did we exterminate them?

* In Asia, the descendants of the "Homo-Ergaster" disappeared, just when WE appeared.

Could their disappearance be attributed to the first campaign of genocide of "Modern Men"?

* Horses are related to zebras, and lions to tigers. However, we are the only species of "Homo."

Did we exterminate all other species of "Homo" (our very distant cousins)?

* At the end of last Ice Age most of the big mammals disappeared from all continents, with the exception of Africa. Until now, the reason was unknown. Now it is attributed to our ancestor ("MODERN MAN" or "CRO-MAGNON" or "HOMO SAPIENS-SAPIENS" or "WE")

* The mammoth in North America disappeared 8,000 years ago, together with many other big mammals, among them the terrible predator, the saber-tooth tiger.

It happened just when the humans arrived in North America?

Was it by chance?

* Radiocarbon studies from the archeological excavations in Monte Verde, south of Chile (South America) indicate the presence of men 12,800 years ago. The humans met there with very favorable environment with many big

animals: mammoth, gigantic sloth, american horse, saber-tooth tiger, etc. The extinction of all these great mammals can be attributed to the climatic changes or what is more probable, to the slaughter committed by modern man (us).

* The great bird "moa" disappeared when the first time humans arrived in New Zealand.

* The list becomes interminable. Where and when WE appeared, the animals disappeared.

Nowadays, the situation is dramatic: there are an enormous amount of species of animals, fish, plants, insects, etc. on the way to extinction. Now the rate of extinction of species is 1000 times bigger than normal.

* The competition in business is another aspect of violence and became ferocious.

There is a constant war (competition) between companies. Which one will be successful? Which one will show the biggest profit? Whatever the cost!

The consequences are: stress, dismissals of the personal and the suffering of everybody, especially their families.

* Other aspects of violence that exists everywhere are assaults, assassinations, terrorism, etc.

* A gene for violent behavior exists?

Latest genetic discoveries confirms, that YES - it exists!

This gene of aggressiveness is our inheritance, which we received through the countless stages of evolution. It is the same gene that allowed us to make possible the spectacular development of our civilization. Without these genes we would still be jumping from one branch of a tree to the other.

However, other effects of these genes are the never-ending wars, atrocities, and exterminations that existed from the very beginnings of "Modern Man." During the long history of humanity, not a single moment existed without a war somewhere.

* How can we eradicate these genes?

The prestigious physicist Stephen Hawking hopes that genetic technology will allow us to eliminate or at least reduce these instincts of aggression. This gene became an "instinct of idiots."

The Darwinist selection works very slowly and to eliminate it from our genetic disposition we need a very long time. This is time we don't have! The only hope is in a genetic technology and the only possibility to survive is to eliminate this *idiotic* gene of violence. From the psychological point of view this gene could be called: subconscious mind or primeval instincts.

* Why and where we have acquired this particular gene of violent behavior during the time of the evolution of "Homo-Sapiens-Sapiens. During some 100 thousand years the humanity lived in extremely harsh conditions. The ice age began around 110 thousand years ago and ended 10 thousand years ago. During this time the temperature dropped various degrees and humans had to adapt themselves to cold and always changing environmental conditions - to prepare clothing, shelters and enough provisions to stay alive. To survive humans had to be intelligent, sometimes violent, to hunt large and dangerous animals like: wooly mammoth, saber tooth tiger, bison, etc. These conditions obliged humans to be resourceful and creative and aggressive.

* Animals that live or lived on the planet earth are or were in a natural state of equilibrium with nature. However, humans...NO!

We move to some territory and reproduce: continue multiplying until the total exhaustion of the natural resources and the only option is to occupy another territory.

Praying for peace by holding hands won't be successful. It's a self-deception only. The only way to eradicate violence is to eradicate it from the minds - it's the only option!

All wars don't start with the first bullet - they began in the minds of the aggressors.

DESIRES...OUR MASTERS

From the psychological point of view, desire is a sentiment of yearning that needs the intervention of will to direct the action for possession or getting the pleasure from something. It's an impulse associated to the idea of an objective.

Desires shape our lives like no other characteristics of our minds.

We think that we control them, but the reality is different: desires are in charge of our lives and we have little control over them. Desires are our masters and we are their slaves!

It would be beneficial for everyone first to analyze desires and then to take action. Therefore, to make this task simpler we can divide desires into three categories:

1) Reasonable desire: for food, housing, clothes and a logical standard of living. However, it depends on the society in which we live. In some societies a car is a necessity and it is impossible to live without it. However, in another society a pair of shoes is an unheard of luxury and only few can afford it. Reasonable desires are flexible and often misunderstood ideas, but there are indispensable of survival of our species.

2) Irrational desire: for nonessentials such as luxuries, extravagances, impressing people, etc. The fulfillment of these desires will not bring happiness. The man wrongly assumes that the possession of some desperately desired object will make him happy and

satisfied forever. However, this is a big mistake: after getting the desired object he immediately needs something new to desire and so it goes, on and on! Our modern age of excesses is based upon these kinds of desires which, by the way, are destroying our civilization by abuse of natural resources, contamination, pollution, etc.

3) Impossible desire: we can never get it, so why waste our time in vain?

We will never reach a star in the sky…On the other hand, maybe we will?

SEXUAL INSTINCT

Two universal instincts rule our lives: the instinct of self-preservation and the instinct of reproduction of the species.

From now on, I will deal with the second one: the instinct of procreation.

Almost all species utilize sex for reproduction.

Why?

The prevailing theory is that sex mixes genes at random so the new beings will be born with modified genes. Some of them will adapt to new environment. Therefore, sex will assure the survival of species...some of them will survive and pass these genes to the next generations. However, many species are not using sex for reproduction, like some species of snakes, insects, fishes, etc. However, these species do not last longer than a few hundred years.

Asexual reproduction is possible, but normally does not occur.

In all species, animals and humans, the character of the female is more peaceful than the male - she does not need to fight with other males for sex. Also, her appearance is different - usually the female is smaller than the male and is less attractive (women are an exception).

Differences between the body and mind are considerable.

What news! However, not all differences are so obvious.

In 99% of the time of evolution, "Homo Sapiens" lived in African forests, in groups of 50 -100 individuals.

Men hunted for animals, fought with other tribes and were looking for a mate (good-looking).

So he had better capacity of orientation and organization: they could situate themselves inside the territory and better understand the maps. They needed these aptitudes for hunting and for war.

The principal occupations of women were searching for nutritious plants, fruits, nuts, etc, taking care of the children, and finding a good mate (good-provider).

Therefore, women had to remember the location of plants and all objects inside the dwelling. It's the reason why women are better observers.

The sexual behavior of women is determined by maternal instinct - to protect her descendants.

A woman is not always faithful, but the reason of infidelity is the preservation of her children, to get help and security from the other partner.

Amazingly, some DNA studies of sexual behavior in marriages, in the occidental societies, showed that approximately 10% of children are not from a legitimate partner. Promiscuity is nothing new and it's common!

The sexual instincts of men and women are similar in intensity: however, the reasons for their behavior are very different.

The sexuality in men consists of a very powerful necessity to disperse his genes as much as possible. To satisfy his instincts, men acts directly, sometimes violently and searches for a partner through eyes - she must be good-looking.

However, what is good-looking? - Good-looking means being healthy and well-prepared for the reproduction. The ideal for a woman's beauty is a big bust and narrow waist. A perfect proportion between the diameter of hip and waist is 0.7.

The medical studies show that women with just these characteristics have better possibilities of getting pregnant and giving birth to healthy children. What is beautiful is also good for reproduction!

The woman pretends to be young and healthy with her clothes, cosmetics, plastic surgery, liposuction, etc.

In a man, the possibility to catch a partner depends on his wealth and his aptitude to be a good provider. A man wants wealth, power and status...a woman wants protection for her offspring and a partner who can offer her security, money and wealth.

In our society each sex displays his/her qualities.

The women display her beauty.

The men show his BMW or his Rolex.

In human species newborn babies are defenseless and need maternal care for a long period of time. Mother alone could not accomplish this task: she needs help of a caring partner.

This point of view is a little cynical.

Why does loves exist?

Love is an instrument of evolution for the preservation of the species, to scatter the genes of individuals which are making love.

Nowadays, in all occidental societies, confusion about women's rights exists. Women belonging to feminist organizations do not believe in separations of duties and demand complete equality.

However, we all are different! Women want equality of privileges, but not of obligations!

Just as a man cannot give birth, a woman cannot take care of her children alone.

All women that work after giving birth endanger the mental and physical health of their newborn children -

they are stressed, unhappy, and cannot accomplish their task of being a good mother.

Lately, communications media criticizes customs and religions of Eastern countries. I consider that we can not criticize the way of life of different cultures, especially when we have nothing better to offer.

In the animal world, the diversity of sexual behavior is just amazing, but it is always to secure the survival of descendants.

Here are some examples of these behaviors:

The female penguin, Adele demands a stone to build a nest from each one of her boyfriends before having sex. Each of her boyfriends thinks that he is the father and so defends newborn-penguins against predators.

The male of a spider, "Red-Back" jumps into the female's mouth, seconds after inseminating her and converts himself into her meal. Because the insemination and later cannibalism take a long time, his sacrifice betters the chance that his sperm will fertilize the female.

Poor male…he has not much confidence in his other half!

The scientists also found out (using DNA) that 54% of baby chimpanzees were not from the legitimate partner of the mother, and the chimpanzees are the species closest to human.

What an incredible variety of sexual habits!

MEMORY – CAN WE DEPEND ON IT?

* From the psychological point of view memory is the ability to store, retain, and recall information. Then, it is the task of the intellect to compare recalled information with current conditions that can be equal, similar, or completely different and draw logical conclusions (unfortunately, not always logical).
This process serves to not repeat an error. This course of action is absolutely necessary for all living organisms to survive.
MEMORY IS A TOOL OF THE INSTINCT OF SELF-PRESERVATION.

* However, memory is not always correct, some memories are a mixture of fact and fantasy and some are completely false.
Recollection from memory can be distorted by external factors: suggestions of parents or counselors, opinions of persons with authority, inconvenient events, etc. or by internal factors such as: imagination, fantasy, wishful-thinking, powerful emotions, psychological trauma or even dreams, which can be taken as being real, etc. However, when emotional shock is too powerful, then a complete memory loss can take place, - known as amnesia. It is a mental strategy to keep the mentality relatively healthy by forgetting traumatic experience.
Then, how can we know if our memories are true?

It is not an easy task. Mostly, we are not aware that our own memory recollection is false. Therefore, we need an outside source to help us become conscious of our errors: somebody, who can correct us and show our mistakes.

*It is an upsetting fact that we can not entirely depend on our own memory.

Therefore, on what can we depend?

To a certain extent we must depend, but we must always be aware that we do not possess the ultimate truth and that we must be tolerant. Other people can have different memory recollections!

* I remember a Japanese movie of a murder, in which each one of the seven witnesses told a completely different version of what really happened and who the assassin was.

* History is a collective memory of a large number of people: nations, empires, kingdoms, monarchies, etc. Lamentably, what we know about historical events is distorted.

It is distorted unintentionally, because of beliefs or of erroneous memories.

It is distorted on purpose, because the history is written by the winners.

However, even with all these shortcomings, history gives us a profound advantage. We can predict the future based on past events.

* How many of my personal memories, especially from the time of war, are false?

Mostly it does not mater, but in any case, I have no possibility to find out. Nobody that I knew then remains alive. Therefore, I cannot compare my experiences with anybody else.

However, I have one "hypothetical" period of amnesia from the time of war, when I was a teenager. I have no proof beside my incredible sensation.

About 30 years after the end of the Second World War, I traveled with my wife to my native city, Warsaw, where I spent five years of my teenage life under German occupation. It was a terrible time - every day we had to suffer assaults, battles, and executions.

One day, I visited friends, family, and places where I lived before. However, in a walk through a wooded park, through which I used to go to school and near the place where my family lived I felt an incredible sensation of horror.

I have no recollection of anything that happened at that place. At that time, I did not go to this park again, but every time I saw these trees in the Saski Park, even from far away, I felt the same sensation of terror.

Then, another 15 years passed and we went to Warsaw again, but I did not forget this dreadful sensation. Of course, one of the first things I did was to go to this park. The same frightening emotion appeared. Therefore, I started to look for the exact place of my terrifying feeling. I started to walk in one direction and then in another. Consequently, I could pinpoint the exact place, which is about five meters away from a very old tree and near the entrance to Saski Park from Marszalkowska Street.

I could not remember anything. However, in this place many battles and executions took place. It was also near the old Jewish ghetto.

Then, I sat nearby on a bench and started to concentrate. Soon, terrible images appeared, but they were so powerful and so terrifying that I could not go on any more and I left this riddle unsolved. I still do not have recollection of what happened there, but I am sure that something did occur.

Was I an observer?

Was I a victim?

Nobody is still alive to tell me the fact.

* The second case of amnesia is a real one, and it is not my imagination.

About 19 years ago, a car accident left me more dead than alive. After a long time I woke up in the hospital. I have no recollection of what happened and there is a gap of a few months prior to this accident that is completely lost from my memory.

I was not even aware that this memory gap was so big, until a relative asked me how I spent Easter at her house. I did not remember anything. Only then did I realize that I could not remember a few previous months - the accident happened after Easter.

* It's surprising, but sometimes memory does not depend on chemical reaction and electrical connections inside the brain. The relationship between the physical body and immaterial mind has been studied for many centuries, but nobody has found an answer to this fascinating question. Maybe there is no answer!

SPORT: AN ESSENTIAL TOOL OF HUMAN DEVELOPMENT

Sport is elemental to human beings. All of us threw a pebble or a ball. All of us enjoyed playing with a friend or running along a track.

All of us, sooner or later, formed teams to compete in games of skill.

Beyond all that, there is a thrill and excitement that comes from races, competition or games.

Games can teach us everything that is useful and indeed can be a tool in human development. Games are something that define us as human beings and something that allow us to explore our common roots wherever we live, regardless of whatever faith we worship.

* Play, sports, and games are changing the lives of people and the destiny of nations as never before.

People enjoyed them always and everywhere: jumping over the bulls in Crete or throwing a rubber ball through a ring in ancient Mexico. Here the similarity ends. In Mexico, the captain of the losing team was "rewarded" by having his head cut off. However, some archeologists sustain that it was the captain of the winning team who was decapitated, as a sacrifice to the gods. In both cases it was not, from our point of view, a pleasant experience.

Now, we must separate sportsmen from spectators.

* Sports, at least in theory, are activities that should keep the body in good physical shape and keep the mind alert and lucid. However, nowadays that is not the case. The

sportsmen risk their lives and their health to win at whatever the cost. Now reins the TRAIN-TIL-YOU-DROP mentality.

It is obvious that many retired athletes are plagued by chronic health problems resulting from overtraining.

Competition is ferocious!

A very long time ago, there was something known as the "spirit of sport." However, now this concept is completely lost. It was about "fair-play" and good behavior. These are now strange ideas! When sports became big business, very big business, when enormous billboards advertising every imaginable gadget covered all the space in stadiums and when prices to TV rights became sky-high, then the essence of sports was lost. Also became forgotten the difference between professionals and amateurs. Sports became big business only - nothing else!

Anybody who can kick a ball, run fast or somebody who is able to do extremely well in any sport, can became an instant millionaire and an idol worshipped by masses.

* For the spectators it's an outlet to discharge their sometimes hidden tendencies for violence, fanatical nationalism and for the most elemental, but frightening instincts.

 Why are there so many hooligans in every country?

* From where does the enormous emotional energy experienced by the spectators and sportsman come from? Sports awake many primordial and powerful human instincts. In the first place, it is an outlet for aggression, as I already mentioned before.

All competitive sports played between two opposite teams are war, as any competition is war, not on a battlefield, but inside a stadium.

* However, that is not all. The history of sports goes back to the beginnings of humanity.

The games, between teams of boys with goal, ball, etc. were preparation for hunting and for war.

Girls played with dolls made from pieces of cloth - the preparation for motherhood.

* However, a new approach of many associations: like human rights, anti-discrimination and many others, is to urge the equality of the sexes. Now women are in the army in combat units, in the air force as fighter pilots, etc. - and have positions not corresponding to their innate abilities and instincts. Many tasks that from the beginnings of humanity belonged to men, now also belong to women.

* There is still another component that makes the games so emotionally stimulating. When humans first became humans, they danced around the fire inside the caves - primitive rites of fertility. Later on, the same dances were transformed into religious festivals: many of them were fertility rites, with the strong influence of magic. Nowadays in many competitive sports the ball must pass through an arc, ring, or any other hole, big or small. The symbolism is obvious! Sports are modern rites of fertility. It's obvious that they are so emotionally charged. When the scientists observed through a microscope for the first time the behavior of spermatozoids, with great surprise they discovered that their actions are identical to football players. They push, run, and want to arrive first...score a goooooooooooooal or in the case of the spermatozoids to impregnate.

The similarity is more than accidental! When, during a soccer match, a player kicks a ball to the back of the net, he enters into a sort of "sports orgasm," shouting like a madman, "GOOOOOOAL!"

*Sports have the power to unite people, and also the power to split them apart.

The Olympics and football and most sporting events became a nationalistic medal count. How many medals each country gained?

It is not about the beauty of movement or the skill of a sportsman, it is only about which country won most medals. Therefore, when sport is played between countries, then nationalism, with its potential for ugly chauvinism, is never far behind.

This nationalistic madness can sometimes have a sinister ending.

* Just one example: The attitude of Argentina's people changed after winning the world soccer championship in 1978? The collective unconsciousness became more nationalistic and self-confident than before.

All wars, including the Falkland War, began not with the first bullet, not because some general or dictator gave an order, but because people's subconscious minds was prepared for war. Games are like everything else. When they are played in moderation, they are a wonderful instrument by which to develop personality, obtain a healthy body and improve society. However, when they become an obsession they destroy a person's mind, his health and the society.

* The Olympic Games and the football association, FIFA instead of being amateur sports became a professional business. The best players became immensely rich and the directors of these "sport" spectacles earn, from salaries and bribes, enormous amount of money. But everything has a price! And the price is an enormous pressure upon sportsmen to win, whatever the cost.

It is a well-known fact that we are living in an extremely competitive world, too much so for my liking. My impression is that we have overdone it!

HOW THE ENVIRONMENT CHANGES THE PERSONALITY?

The environmental conditions have a decisive role in shaping human's physical bodies, as well as their mentalities. For instance, people who live in a cold climate have whiter skin than people that live in tropical environment - where the skin takes a darker color to protect the body against solar radiation.

One of the best examples is in Europe.

The mentalities of the Northern people are "time-conscious," and the mentalities of the people from the south of Europe are "timeless."

What does that mean?

Both of them are poles apart and there is no understanding between people in Europe between North and South nations. As always, to look for the solution, we must look into the past.

* How did the "time-conscious" mentality evolved?

To survive the cold winter's people were obliged to accumulate food and fuel for the ice-cold climate. Organization of deposits had to be perfect and everything had to be on time - in this case, before winter. Provisions had to be stored, wood accumulated and clothing prepared. Everything had to be ready, organized and on time!

The mentality of these people had to be well organized. They were always looking ahead, making plans and trying to foresee the future.

They don't live for today. Their minds are always projected into the future or remembering the past, so as not to repeat errors committed previously. They are extremely punctual - sometimes even slaves of time, with narrow-minded, intolerant and materialistic minds. They live a stressful life, worrying about what the future will bring and can not easily express their feelings. They are serious and with little sense of humor. However, they are men of action, warriors, and conquerors. They usually get what they want, but they always want more and more, without any limits.

They are loyal to their societies. Cooperation of the whole society was and still is, essential for survival. Nobody would be able to survive in these extremely difficult environments alone.

The nations in which they live are successful, well-organized, uncorrupt and rich - so-called "developed nations!"

However, "time-less" people lived in a warmer climate and friendlier environment: therefore, their minds evolved very differently.

Surviving was much simpler!

To survive, it was not necessary to foresee the future, neither to organize deposits nor to accumulate provisions for the winter. Therefore, the help of the society was not necessary.

They usually consider the government to be a hindrance. Therefore, tax evasion as well as corruption is common. Nepotism and corruption are usually associated with the developing countries.

However, they are loyal to their families.

"Time-less" people are not punctual. Instead of organizing, they improvise. Their minds are usually chaotic, extroverted, and sociable and they enjoy good living. They like dancing and singing and are happy-go-lucky and carefree.

Live NOW… they are not concerned about the past or about the future.

Of course, pure "time-less" or pure "time-conscious" minds do not exist - only a prevailing tendency - to live for today or to live for the future!

However, there is one riddle inherent in this theory.

Normally, people who live in a cold or mountainous environment are short and corpulent. However, people in the northern part of Europe are tall and slim. How is that possible?

Their physical bodies must have evolved in a hot climate, where tall bodies are better adapted to high temperature, as less body surface was exposed to solar radiation. Short and corpulent bodies are better adapted to a cold climate, as a smaller amount of body surface is exposed to cold.

The only explanation to this apparent enigma is that their bodies evolved before their minds evolved…

Therefore, a long time ago, the tribes from Northern Europe immigrated from a hot climate in Africa, where their bodies had evolved, to a cold climate in Northern Europe, where their minds evolved.

However, this is merely my hypothesis and there is no proof of the veracity of this supposition.

COLLECTIVE UNCONSCIOUS
(Referred to as CU)

The CU is a term coined by a psychologist Dr. Carl Jung to describe a part of the unconscious mind, shared by the society, people or all humankind. It's a product of ancestral experiences. The CU does not develop individually but is inherited. It consists of pre- existent forms, the archetypes, universal mental predispositions not grounded in experience, impulses of which a person is not aware, such as archetypes, universal images, ideas, etc. The CU is a foundation of human unity.

There is also a "group unconscious" for every group of people and culture (family, tribe, ethnic group, nations, military unit, even sport club, etc.)

Like all other characteristics of our minds, CU is an essential tool of evolution. When a group of early humans went hunting or fighting with another tribe, they needed to have some mental communication between all of their members to devise strategies. For that, the CU is crucial. The possibility of survival or of a successful hunt is much better for a group of men than if each one acted alone. That is the reason that CU is much stronger between men than women and also the reason that very few women are hooligans, supporters of a football club or belong to a gang.

It is extremely difficult to recognize the existence of the collective unconscious. It is like the air we breathe: we know that it exists, but we are not conscious of it.

The only way to recognize CU is by comparing our beliefs, our values and our mentality with those of other cultures, by observing them impartially and without judgment. Unfortunately, our mentality judges and condemns everything that differs. It is much easier to judge than to understand! However, when we are always in accord with the society.... beware! Our CU is thinking for us!

The CU acts by impulse or emotion, has neither mercy nor compassion and is independent of the mentality of the people that integrate it...it can even be just the opposite!

A group of hooligans on a soccer field is the best example: they can be violent and wild. I can even imagine a group of university professors in a stadium even they could turn into beasts!

All humans possess a typical CU which changes with time. (*It would be very interesting to know the mental characteristics of our cousins, the Neanderthals, though already extinguished, to see in what way they were different from us*).

How did our ancestors think?

I will give some examples:

Our ancestors thought differently, they had other values and ideas. They were building enormous cathedrals during the Middle Ages in spite that they knew that it would take hundreds of years to finish them, and would require the enormous effort of many future generations.

Who now would build skyscrapers to be completed after 200 years?

They were convinced that it was worth the trouble: they did it for God.

The CU in Europe changed with the arrival of the Renaissance. They stopped building cathedrals and committed themselves to science and art.

*In the same way that a person can be mentally sick: schizophrenia, psychosis, dementia, etc, also the whole society also can be mentally ill. Nowadays the CU of our Western civilization is self-destructive.

Motion pictures are violent, horrifying, and morbid. Music is aggressive, loud, and deafening. The paintings represent horrible and preposterous scenes. Compared to the last century, the literature is practically non-existent. When I see new art, which is an expression of sub consciousness, I have many doubts about the future of humanity.

* The CU of the North American society entered into a very dangerous point. Nationalism reached a perilous stage and obsession with the terrorism is auto destructive. They spent enormous amount of money on new weapons. They are completely brainwashed through different mediums of communication - patriotic fervor is extreme - flags are hanging everywhere.

I have seen the same during WWII in Germany: flags were hanging and military marches were played all the time.

Will the Americans need some disaster to wake up, like the Germans did?

*Some years ago, I heard a curious theory that approximately every six centuries the CU changes: trends, tendencies, objectives, and the course of history of all people on our planet. Studying history, I have found this theory to be surprisingly true.

Some examples:

In the twelfth century BC: - the Pharaoh Akhenaton introduced the first monotheist religion in Egypt - the end of the Minoa civilization in Crete - the Trojan War - the invasion of India by the Indo-European tribes.

In the fifth century BC: - the Greek philosophers - Buddhism in India - Confucius and Lao-Tsu in China - Zoroaster in Persia.

In the first century AD: - The beginnings of Christianity - Roman writers and philosophers - the utmost expansion of the Roman Empire.

In the seventh and eighth century AD - Mohammed was born and the beginning of Islamic religion. - The collapse of the Roman Civilization – Beginnings of The Middle Ages - city of Teotihuacán in Mexico was abandoned – the Mayan's cities were deserted.

In the fifteenth century AD: - the end of the Medieval age - the beginnings of the Renaissance - the discovery of America - Great painters, scientists, and conquerors such as Leonardo Da Vinci, Michael Angel, Copernicus, Galileo, Newton, Columbus, etc.

From the fifteenth century, the CU began changing faster. In the nineteenth century AD: - the end of the Renaissance - the beginning of the Industrial Age.

Great writers: Goethe, Schiller, Dostoyevsky, Tolstoy, Oscar Wilde, Victor Hugo, Tennyson, etc.

Great composers: Beethoven, Brahms, Puccini, Verdi, Hayden, Tchaikovsky, Chopin, Wagner, etc.

Great scientists: Darwin, Edison, Koch, Freud, Max Planck, Einstein, etc.

During the nineteenth century the foundations of almost all scientific discoveries were established, which were applied later on in the twentieth century: the automobile, airplane, radio, television, even the atomic bomb, etc. During the nineteenth century, the CU became more materialistic. When the biggest desire became to possess as much as possible, even sacrificing happiness and health! This way of thinking is illogical and absurd, and leads to greed, extreme competitiveness, snobbery, infelicity, anxiety, stress, etc.

These century brought: the end of feudal epoch - beginnings of nationalism - the formation of states, such as Germany, Italy, etc. (previously on these territories were many small kingdoms).

Nationalism is a scourge of our age: when millions of young men are sacrificed during the never-ending wars.

During the twentieth century AD - there were few great writers, composers, or painters – but sciences and technology developed spectacularly.

I can remember that during my youth, some 80 years ago, there were not so many new gadgets, and people thought differently, had other values and other ideas. During my lifetime, CU changed radically! It is evident that now CU is changing even faster; it varies every few decades.

Why is the collective unconscious changing?

CONSCIENCE AND THE SENSE OF RIGHT AND WRONG

What is conscience?

In philosophy, it means a mental characteristic that distinguish between good and bad.

We do not know if this mental characteristic exists only in humans. It is possible that it exists in some animal's especially big apes, but naturally in smaller degree only.

When we must make a decision, it is essential to be fully conscious and wide awake. The conscience acts as a referee, and decides which action, between all others possible, should be taken. This "referee" is the CONSCIENCE.

I will give a simple example:

A man is attacked by a lion and has the following options:

a) The instinct of survival orders him to escape. It is an instinctive action caused by fear.

b) Nearby is a tree: reason suggests that he should climb the tree and escape from the lion. He remembered that in a previous occasion he had saved his life in the same way.

c) His son is nearby and in danger of being attacked by a lion. His social duty and love for his son orders him to defend his son (a moral law). He has two options: to escape and save his own life, or to fight against the lion, defending his son and sacrificing his own life.

His conscience must decide which action he should take.

This is a simple example; however, during our lives we are obliged to make decisions that cause stress - now so common in our society.

When we must make a decision: in our daily lives, our professions or our social lives, we do not know how to act. We need some rules to guide us to choose the best alternative. Thus appeared the concept of good and bad!

To act correctly, is to act well.

What does it mean?

Good means: to obtain the maximum wellbeing for oneself, wife, children, parents, and other relatives, and for the society in general, such as one's nation, clan, tribe, or even football club to which he belongs.

The nature of a man is to act according to his basic instincts: survival and reproduction. Nevertheless, he lives within a society, and the society needs all members to behave according to its laws which assure the stability and social order. However, human nature is purely instinctive and in many circumstances, it's the opposite to the laws of society.

This is the biggest conflict of human beings!

Probably, the biggest conflicts are caused by love and matrimony, both of which are very important to the stability of the family and of society.

When an individual must choose between his sexual instinct and his moral obligation (marriage), in most cases the instinct prevails. Instincts are much stronger than any laws, obligations, or duties.

It is exactly the same conflict that gives unlimited subjects to writers, poets, films, theatrical dramas, and soap operas; it is also the same conflict that provides a very high standard of living for lawyers, judges, priests, writers, psychologists, etc.

To assure the stability of society the humanity had elaborate written laws, to which all members are forced to obey: like the code of Hammurabi (probably the oldest), the Roman laws, Napoleons, and many others.

Nowadays, an enormous amount of new laws, decrees, etc…appear every day and in all countries.

All religions teach us to behave ethically and morally: aim is to protect the society.

In Christianity, the Ten Commandments - the first two steps of Patanjali yoga, the eight steps of Buddhism, etc.

All these commandments are rules for humans to live in harmony with society and to behave according to the laws that exist in these societies. All these rules assure personal survival and the survival of society.

The concept of good and bad is not universal, it changes with time and differs in each society - it is not absolute: It's relative!

The modern way of life allows us to travels to distant lands and to communicate with other people. That gives us the possibility of observing other customs and mentalities and other rules of good and bad. Sometimes, we may not like them. However, they exist and we must unavoidably accept them!

The intolerant, ignorant, and closed minds are the consequence of ignorance of not accepting that simple truth, that other people have the same rights as we have!

In my native Poland, there is a short story by a famous writer.

Someone asks a small boy in Africa, "Do you know what is good and what is bad?"

He answers, "of course I know; bad is when somebody steals my cow."

Very well, "and what is good?"

"Good is when I steal the cow," he answers.

This small anecdote is an example, although exaggerated, that the rules of behavior are valid within one society only - they are not universal!

With the bigger population, these laws are getting more complex. Thus, we have laws, judges, attorneys, police, jails, etc.

On the other hand, a big part of the society lives in the shadow of bad: the thieves, assassins, corrupt politicians, etc.

EMOTION

In psychology and philosophy, emotion is the generic term for subjective, conscious experience that is characterized by psycho physiological expressions, biological reactions and mental states. Emotion is often associated with mood, temperament, personality, disposition and motivation, as well influenced by hormones and neurotransmitters such as dopamine, noradrenalin, serotonin, oxytocin and cortisol.

Emotion is often the driving force behind motivation: positive or negative. For example, the experience of fear occurs in response to a thread.

Emotion can be differentiated within the field of affective neuroscience:

* Feelings are best understood as a subjective representation of emotion, private to the individual experiencing them.

* Moods are diffuse affective states that generally last for longer duration than emotions and are usually less intensive than emotions.

* Affect is an encompassing term, used to describe the topics of emotion, feeling and moods together, even though it is commonly used interchangeably with emotion.

* Emotions operate on many levels: physical and psychological.

Emotions bridge thoughts, feeling and action – they

operate in every part of a person, and the person affects many aspects of the emotions.

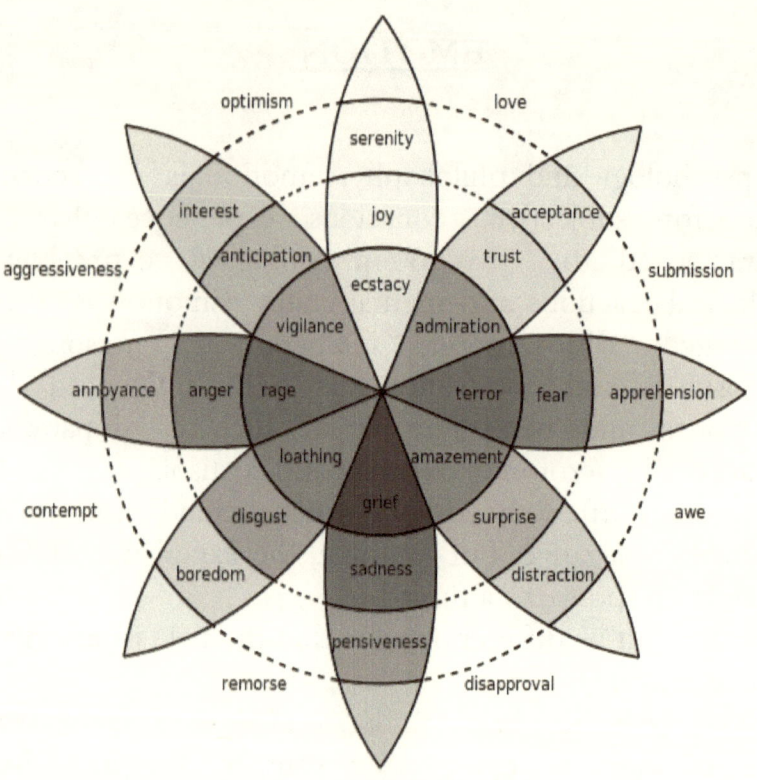

How can we cope with emotions?
* People, who ignore, dismiss or repress their emotions, are setting themselves up for physical illness, including cancer, arthritis and many types of chronic illnesses. Negative emotions such as fear, anxiety, negativity, frustration and depression cause chemical reactions in the body that are very different from the chemicals released when you feel positive emotions such as happy, content, loved and accepted.
* Underling much of our behavior is a belief system. This system within us filters what we see and hear, affecting

how we behave in our daily life. For example, a child raised by an angry man or woman will view people with belief that anger is bad or that it is something to fear. Another example would be someone who is intelligent but has never been encouraged or honored for his intelligence, this person might belief that he is stupid.

It takes a lot of work to look at yourself and identify the beliefs that are affecting your life in a negative manner. However, knowing your beliefs will give you a sound basis for emotional freedom. It's wise to deal with the beliefs before dealing with the identification and release of emotions. The only person who can change feelings is you. A new relationship, a new car, a new job, this things can momentarily distract you from your feelings, but no other person, no material possession, no activity can remove, release or change your feelings.

* There are two basic emotions: love and fear. All other emotions are variation of these two emotions. Anxiety, anger, sadness, depression, inadequacy, confusion, loneliness, guild, shame, are all fear-based emotions. However, joy, happiness, caring, trust, compassion, contentment and satisfaction, are love-based emotions.

* Emotions have a direct effect on your health. Fear-based emotions stimulate the release of one set of chemicals, while love-based emotions release different chemicals. If the fear-based emotions are long term they damage the immune system, the endocrine system and all other systems in the body.

* It's impossible to change or control the emotions. However it is possible to accept them, living peacefully with them, releasing them, even managing them, but never control them. Some people seem to live a normal life, but suddenly they explode in anger at something

trivial and harmless. It is a sign that they try to control or repress their emotions.

* Confusion occurs when people are trying to get to know their emotions because they speak in general terms rather than specific emotions. A good example of this behavior is depression. You may be experiencing loneliness, boredom and lack of creativity. You may be feeling abandoned because of death or divorce. If you just say you are depressed you will have great difficulty releasing your emotions or finding a solution. A good example is the difference between jealousy and envy. Jealousy relates to being resentful of a person's advantages: social position, education, profession or it can refer to a rival in love or affection. Envy is a discontentment or resentment aroused by anther's success or luck.

* When we have an experience that we find painful or difficult, and are either unable to cope with pain or afraid of it, we dismiss this emotion and either get busy, exercise more, drink more, eat more or pretend it has not happened. By doing that we do not feel the emotion and result is repressed, suppressed or buried emotion. Emotions that are buried for a long time normally cause physical or psychological illness.

 Here are a few examples of the methods people use to avoid feeling their emotions.

* Ignoring feelings
* Pretending that something did not happened
* Overeating
* Eating foods with too much sugar and fat
* Excessive drinking of alcohol
* Use of drugs
* Using prescription drugs such as Prozac or tranquilizers
* Exercising compulsively
* Compulsive behavior

* Excessive sex
* Always keeping busy
* Analyzing continuously
* Working Excessively
* Keeping superficial conversation
* Burying angry emotions under the mask of peace and love
* When you have repressed emotion, your behavior and reactions to events in the present moment are really reactions to past events as well as the present ones. This has negative effects on all relationships in your life. You can not be fully present with those you love.
* Don't be afraid of your emotions. Don't fight them, run away from them, blocking them out. Welcome them; be with them, regardless of what they are. We were born with all emotions, they are neither good nor bad - they just are. Emotions dissipate and slowly disappear if you feel them, and are with them. Just close your eyes and feel them as deeply as you can.

CONCLUSION

The subject of the mind is never-ending. However, I have to stop sometime. Therefore, I decided to publish what I already have written.

All minds' characteristic and also of the body were fixed because an enormous number of events took place before:

1) Because the first living cell "emerged" on Earth, more than 3.500,000 years ago and evolved into all creatures which are living today. This cell is our first and our oldest ancestor.

2) Because an asteroid plunged into the Earth 65 million years ago and destroyed the dinosaurs, and therefore allowed the evolution of mammals.

3) Because five or six million years ago, the climate change in Africa transformed the tropical jungles into savanna, what obliged our ancestors to walk on two legs and not to jump from one branch of a tree to our another.

4) Because our ancestors around 2.5 million years ago learned to control the fire.

5) Because our ancestors, some two million years ago, began to manufacture stone tools.

6) Because 150 to 200 thousand years ago, somewhere in Africa, evolved our direct ancestor, the modern man (Homo-Sapiens-Sapiens).

7) Because more than 100 thousand years ago, the mind of the modern man evolved to self-consciousness.

8) Because more than 50 thousand years ago, the humans invented spoken languages.

9) Because some 10 thousand years ago, the end of the last glacial age permitted the invention of agriculture, and later on, of urban civilizations.

10) Because Alexander the Great conquered Asia.

11) Because the Roman Empire conquered the Mediterranean world.

12) Because Cleopatra had a romance with Caesar.

13) Because Galileo, Newton, and Copernicus revolutionized the concept of the universe.

14) Because during the nineteenth century, scientific discoveries prepared the Industrial Revolution.

15) Because the French Revolution troubled the civilized world with ideas of freedom, democracy, and social justice.

16) Because Hitler invaded Poland and started the second World War.

We are the effect of what already happened and are the cause of what will happen in the future.

From one side, our entire personalities are the results of all our mental characteristics (hereditary and acquired) and from the other side is the consequence of an infinite numbers of events that took place before us.

There is very little place for a real free will!

It's a rather sad conclusion!

We are like puppets dancing on the stage of life, where the destiny pulls all the strings.

FROM THE SAME AUTHOR:

RELATOS Y LEYENDAS "Antigualas de Catamarca".
Spanish edition.
Edited 1984 - "Club del Libro Cívico".

MY TRAVELS THROUGH CALCHAQUIES VALLEYS.
Bilingual edition - English & Spanish.
Edited 2005 - El aleph.

THE COLLAPSE OF MODERN WAY OF LIFE.
Bilingual edition - English & Spanish.
Edited 2006 - El aleph.

WHO ARE THE JEWISH PEOPLE?
Bilingual edition – English & Spanish.
Edited 2007

STORIES FROM MY LIFE.
Bilingual edition – English & Spanish.
Edited 2008 - El aleph.

MYTHS AND FACTS.
English edition.
Edited 2010 Lulu.com.

MISTERIOS DE LA MENTE
Spanish edition.
Edited 2010 Lulu.com

MYSTERIES OF THE MIND
English edition.
Edited 2010 Lulu.com

PERSONAJE INOLVIDABLE
Spanish edition.
Edited 2010 Lulu.com

RECUERDOS DE TIEMPOS DE ÑAUPA
Spanish edition.
Edited 2011 Lulu.com

STORIES FROM BYGONE TIMES
English edition.
Edited 2011 Lulu.com

THE COLLPSE OF WESTERN CIVILIZATION
English edition.
Edited 2011 Lulu.com

WHO WAS IN AMERICA BEFORE COLUMBUS
English edition.
Edited 2013 Lulu.com

THE STORY OF US
English edition.
Edited 2013 Lulu.com